I0484677

Modbus for Field Technicians

Revision 1.0

Copyright Notice

Mailing Address: 3495 Cambie St, # 211, Vancouver, BC , Canada, V5Z 4R3

Thanks to Liz Lucica for all your work in putting this booklet together.

Modbus is a registered trademark of Modicon.

TABLE OF CONTENTS

MODBUS - Introduction 5

1.   There are 4 types of data. 6

2.   There are (were) a Max of 9999 points of each data type. 8

3.   5 Digit vs 6 Digit Addressing 9

4.   What about Scaling in Modbus 12

5.   Floating Point Numbers in Modbus 13

6.   Byte/Word Order – An ambiguous nightmare 14

7.   Bit Order – Sometimes it's a problem too. 16

8.   Modbus and Gateways 17

9.   What about errors / exceptions. 18

10. There can only be one master on a Modbus Serial Trunk 20

11. Multiple Clients of a Modbus slave 21

12. Old device – slow processors – limited capability 27

13. Modbus Ascii, JBUS, Enron and other Variants 27

Modbus RS232, RS485 and TCP/IP 29

14. How Modbus is Transported 30

15. Modbus on RS232 31

16. Modbus on RS485 32

Modbus Resources, Testing and Trouble Shooting 45

17. What to take to site with you 46

18. Trouble Shooting Modbus TCP/IP 51

     Required tools 51

     How to Capture with Wireshark 52

     Capture Filters 57

     Display Filtering 59

     Searching 59

19. Using the CAS Modbus Scanner 61

20.   Converting Modbus 16 bit numbers to 32 bit numbers.................................. 66

21.   How Real (Floating Point) and 32-bit Data is Encoded in Modbus RTU Messages
       69

       The Importance of Byte Order 69

       Determining Byte Order.. 71

       Practical Help 73

22.   Hubs vs Switches – Using Wireshark to sniff network packets 76

MODBUS - INTRODUCTION

Because it is so commonly used, because it is so limited, because some vendors went to a lot of trouble and because some vendors hired bad programmers, Modbus, as simple as it seems, can offer lots of complications.

Modbus was invented to transfer data as well as to program/configure PLC's. For the purposes of this article, we are only interested in the data transfer functions.

1. THERE ARE 4 TYPES OF DATA

Holding Registers

An area of 16 bit words. Intended as read / write. Originally used as programmer scratch pad area and for analog outputs in old Modicon PLC's.

Also known as 4xxxx registers (xxxx is the place holder for the specific holding register's point number).

Input Registers

Think Analog inputs. 16 bit words.

Also known as 3xxxx registers (xxxx is the place holder for the specific input register's point number).

Inputs

Think Binary inputs.

Also known as Inputs.

Also known as 1xxxx inputs (xxxx is the place holder for the specific input's point number).

Coils

Think Binary outputs. Named coils after the coil in a relay which is activated to energize a circuit. The original PLC's were relay replacement machines.

Also known as Outputs.

Also known as 0xxxx inputs (xxxx is the place holder for the specific input's point number).

2. THERE ARE (WERE) A MAX OF 9999 POINTS OF EACH DATA TYPE

When Modbus was invented they thought 9,999 items of each memory type were enough.

Most vendors ignore this limit today – they make clients that can read more and they make devices which can serve more if required.

Older clients cannot poll for more than 9,999 items.

Even though 9,999 was an arbitrary choice there is a practical limit imposed by the protocol. The Modbus message uses a 16 bit word to identify the point number to be read/written. The largest number that can fit in 16 bits is 65535 and hence the highest point number that can be read is point 65535. Most vendors, these days, allow their software to read any points in this range.

400001, 400002 … 409999….. We call this five digit addressing.

So now we come to a naming problem.

3. 5 DIGIT VS 6 DIGIT ADDRESSING

If 40001 is the 1st, 40002 the 2nd ....   We get to 49,999 and then what? 50,000? No!

We introduce an extra zero.

Instead of 40001 we talk about 400001, 40002 becomes 4000002

Thus

400001, 400002 ... 409999, 410000, 410001 ..... We call this six digit addressing.

There are 4 types of data - They are ambiguously identified.

When Modbus was defined, the inventors gave name and identifiers to each data point in each of the 4 memory areas. Each point was given a public and a hidden identifier. When these two get confused so do we.

Holding registers are most commonly identified as

   40001

   40002

   40003

   Etc

The '4' indicated 'Holding Register'.

The remainder of the number is the 'Holding Register' number.

40001 means the 1st Holding Register.

40002 means the 2nd Holding Register.

BUT HERE IS THE IMPORTANT PART

Let's say you want to read, for example, the value of holding register named 40010.

Our intuition expects a Modbus poll to say "Read holding register # 40010".

However Modbus has its quirks. When Modbus reads it sends a message saying "Read Holding Registers - offset from the 1st holding register by 9".

Thus privately (inside the Modbus message) the holding register 40010 is identified as 9.

Example:

Configure your client to read 40108 (Public address)

Inside the Modbus message sent you will find.

Here is an example of a request to read registers

40108–40110 from slave device 17:

Example Field Name (Hex)			
Slave Address		11	
Function		03	
Starting Address	Hi	00	
Starting Address	Lo	6B	
No. of Points Hi	00		
No. of Points Lo	01		

6B(hex)=107(Decimal)

Modbus Message =
Read Holding Register (Function=3) offset by 107 from the 1st holding register. I.e. register 40108.

The same discussion applies to the other data types.

Publicly we number them from 1. Privately (inside the messages) we number them by their offset from the 1st one (i.e. we number the 1st one as zero.)

Another Factor

Some Vendors do not use the 0xxxx, 1xxxx, 3xxxx, 4xxxx notations when itemizing data points.

In the example below the Vendor doc doesn't tell you if it's a holding register or input register and they are numbered from 1. You would have to check the assumption that point number 1 is 40001.

Veris Industries, LLC

H8238/MCM MODBUS POINT MAP

```
F O R M A T
Tnt.    Floar.     R/W    NV     Description
1       257/258    R/W    NV     Energy Consumption, kWh, Low word integer
2       259/260    R/W    NV     Energy Consumption, kWh, High-word integer

                                 Both 257/258 and 259/260 have the same floating point
                                 value.

3       261/262    R             Real Power, kW
4       263/264    R             Reactive Power, KVAR
5       265/266    R             Apparent Power, KVA
6       267/268    R             Total Power Factor
7       269/270    R             Voltage, L-L, avg of 3 phases
```

4. WHAT ABOUT SCALING IN MODBUS

Modbus does not provide a method for transporting large or Floating Point numbers or a mechanism for scaling analog values.

A 16 bit word can only contain values in the range 0-65535. Only whole numbers are permitted.

To work around this many server device manufacturers use multipliers and document them in their manuals. For example, to report a temperature of 58.5 the device reports a value of 585, and makes a note in the manual that the master should scale by 10.

This scaling is achieved by adopting a convention between the client and the server.

What about large numbers > 65535

Modbus does not provide a mechanism but 3 important schemes are widely used.

Long Integers – Two consecutive 16 bit words are interpreted as a 32 bit long integer.

MK10 values – Two consecutive words are used. The 1st reports the number of units and the 2nd reports the number of 10,000's.

Floating Point Numbers – Two consecutive words are used and a scheme. (See section X)

These schemes are conventions and not all servers or clients support them.

The protocol does not identify these big numbers. Only the vendor docs do. What we mean by this is – if you look at the byte stream in a Modbus

message there is no way of telling whether you are looking at two consecutive 16 bit words, or two consecutive words that should be interpreted as floating point, long or MK10 formats. Because of this you always have to look to the vendor docs.

Read more in Appendix 3.

5. FLOATING POINT NUMBERS IN MODBUS

Modbus was not designed to transport floating point numbers. After the protocol was released and in use – some people came up with a scheme to using two consecutive 16 bit registers to transport one floating point number. The scheme is essentially a set of rules for interpreting the bits in the 2x registers as the elements of a floating point number (like a mini protocol). Other people came up with other schemes.

One of these schemes has come to dominate. It is called standard IEEE754.

Some devices (servers) do not support floating point numbers.

Many clients (masters) do not support floating point numbers.

A master and a server must use the same floating point scheme to work together.

Read more in Appendix 4.

6. BYTE/WORD ORDER – AN AMBIGUOUS NIGHTMARE

It takes two bytes to make a 16 bit word. These bytes can be arranged in two ways.

When floating point, long integer or MK10 value is transported there are 4 bytes in two words. The order in which the words are sent as well as the order in which the bytes are packed into each word can change from device to device.

How did this stupid situation come to be?

Some microprocessors arrange the bytes in a word in one order and other microprocessors do it in the opposite order. Some programmers account for this and take steps for the device to serve its bytes in the standard order but some manufacturers had bad programmers who did not care and their device put out data in the wrong order.

Most often you will learn of this issue the hard way – the most common symptom – the values you see in the client are not what you expect.

The jargon word for the order in which bytes are packed into a word is 'Endianess'.

Here is an example of how this works.

Each block represents one byte. The two bytes make a word. The value in each block is in decimal.

1	2

This can be interpreted as

1x256 + 2 = 258 (High Order or Most Significant Byte 1st)

1 + 2x256 = 513

This is ambiguous. Here is how you resolve this –

- Apply common sense - Which value is correct.
- Read the manual and look for the word Endianess' or 'Byte Order'. Some examples are provided below.
- Make an assumption – The protocol spec requires the high order byte to be transmitted 1st so assume it is.
- If your client / master allows, use a function to swap the byte order.

These two FieldServer functions combine two 16 bits words using the IEEE754 rule and make a floating point number. There are two functions because they use the words in different orders.

- 2.i16-1.float-sw
- 2.i16-1.float

COMMUNICATIONS

The protocol to be used is Modbus RTU. This is implemented by the master (PC, PLC, etc.) issuing a poll to the slave (BurnerLogix) and the slave responding with the appropriate message.

MESSAGE FORMAT

DST	FNC	ADR HI	ADR LO	DAT HI	DAT LO	CRC LO	CRC HI

DST refers to the logical address of the slave

Extract from a Manual.

Show High order or Most Significant Byte is transmitted 1st. This is how the spec requires a vendor to serve data.

FNC 03 is a read request and FNC 06 is a write request.
...ber number of the data being requested.
...are mapped as HOLDING REGISTERS. Register addresses
...address 00.
...d. A word is an integer consisting of 2 bytes
as follows:

DST	FNC	DRC	DATA... HI/Lo Format	CRC LO	CRC HI

7. BIT ORDER – SOMETIMES IT'S A PROBLEM TOO

In older Modicon PLC's bits were numbered 1-16. All modern systems use 0-15.

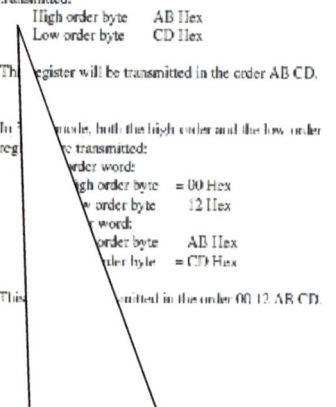

2.3 DESCRIPTION OF THE MODBUS PACKET STRUCTURE.

Every Modbus packet consists of four fields:

1) The Address Field
2) The Function Field
3) The Data Field
4) The Error Check Field (Checksum)

2.3.1 ADDRESS FIELD

The address field is 1-byte long and identifies which slave device the packet is for. Valid addresses range between 1 and 247. The slave device whose address matches the value in this field will perform the command specified in the packet.

2.3.2 FUNCTION FIELD

The function field is 1-byte long and tells the addressed slave which function to perform. The Modbus functions supported by the 3710 ACM are listed in Figure 2.1

2.3.3 DATA FIELD

The Data Field varies in length depending on whether the message is a request or a response packet. This field typically contains information required by the slave device to perform the command specified in a request packet or data being passed back by the slave device in a response packet.

In general, data in this field are contained in either 16-bit or 32-bit registers. In 16-bit mode, registers are transmitted in the order of high-order byte first, low order byte second. In 32-bit mode, registers are transmitted in the order of high-order word first, low-order word second. For example, a 3710 ACM real-time parameter has the content 0012ABCD Hex.

In 16-bit mode, only the low-order register is transmitted:

High order byte	AB Hex
Low order byte	CD Hex

This register will be transmitted in the order AB CD.

In 32-bit mode, both the high order and the low order register are transmitted:

High order word:
High order byte = 00 Hex
Low order byte 12 Hex
Low order word:
High order byte AB Hex
Low order byte = CD Hex

This is transmitted in the order 00 12 AB CD.

Notes in a Vendor manual indicate byte order. In this case, high order byte first thus this vendor meets the Modbus Spec.

8. MODBUS AND GATEWAYS

A gateway is a device that makes data read using one protocol available using another protocol. For example you could read Modbus data from a power meter and serve that data using BACnet to a Building Automation System.

What data must the gateway report if the Modbus is offline or the data cannot be read? It can report the last value read. How old is that value? In this example, we can exploit a property of each BACnet data object called Reliability. When the validity of the data is unknown, like when a field device is offline, we mark the BACnet objects as unreliable. Now a consumer of that data has enough information – he knows the value and if it is reliable. It is his call whether to use the data or not.

Modbus does not have an equivalent mechanism. If a gateway is doing the opposite, for example, reading BACnet data and serving that data using Modbus. If the BACnet link is broken the data validity is questionable. However, in Modbus there is no way of reporting this.

The gateway can take one of two actions – serve the invalid data or – not serve the data – by not responding to the poll. This is the strategy FieldServer gateways use. If the Data is invalid, the Gateway does not respond to a request for that data and allows the client to time out.

9. WHAT ABOUT ERRORS / EXCEPTIONS

Modbus has a limited way of reporting errors. A server / slave device can respond to a message in a way that reports an error. These are called exception messages.

If you are looking at a message byte stream, exceptions are easy to identify.

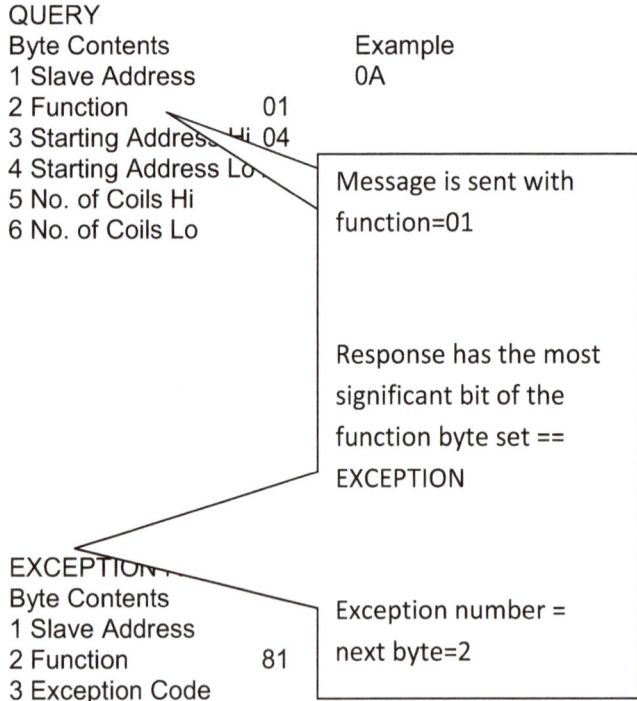

QUERY
Byte Contents   Example
1 Slave Address   0A
2 Function   01
3 Starting Address Hi 04
4 Starting Address Lo   Message is sent with
5 No. of Coils Hi   function=01
6 No. of Coils Lo

Response has the most
significant bit of the
function byte set ==
EXCEPTION

EXCEPTION
Byte Contents   Exception number =
1 Slave Address   next byte=2
2 Function   81
3 Exception Code

Code   Name / Meaning

1   ILLEGAL FUNCTION
The function code received in the query is not an allowable action for the slave. If a Poll Program Complete command was issued, this code indicates that no program function preceded it.

2   ILLEGAL DATA ADDRESS
The data address received in the query is not an allowable address for the slave.

3   ILLEGAL DATA VALUE
A value contained in the query data field is not an allowable value for the slave.

4   SLAVE DEVICE FAILURE
An unrecoverable error occurred while the slave was attempting to perform the requested action.

5   ACKNOWLEDGE
The slave has accepted the request and is processing it, but a long duration of time will be required to do so. This response is returned to prevent a timeout error from occurring in the master. The master can next issue a Poll Program Complete message to determine if processing is completed.

6   SLAVE DEVICE BUSY
The slave is engaged in processing a long–duration program command. The master should retransmit the message later when the slave is free.

10. THERE CAN ONLY BE ONE MASTER ON A MODBUS SERIAL TRUNK

Modbus is a poll-response type of protocol. A master issues a message. If the address in the message matches the address of a server device it will respond (if it can). All other devices remain quiet all the time until they are sent a message with a matching address. The master must wait long enough to process the response before sending the next message. If it doesn't then its next poll and the response from the previous may overlap.

When Modbus over Ethernet is used, more than one master can poll a server device for data. The number of queries that a server can process simultaneously is dependent on several factors – does the vendor support multiple simultaneous socket connections and how many do they allow. Vendors hardly ever publish this information.

11. MULTIPLE CLIENTS OF A MODBUS SLAVE

We are frequently asked how you deal with a situation where you have more than one client for a slave(s). The Modbus spec does not support this but we have a solution.

The essence of the solution is to use a multi-port FieldServer. Connect each client to its own port and the slave(s) to their own ports. Each client will see a single virtual slave(s) on its network. This not only solves the problem but is extremely efficient. Of course the FieldServer needs to be correctly configured.

In a situation like this we exploit the FieldServer technology known as 'Port Expansion'.

Figure 1: Normally it is not possible to connect two clients to the same slave. There are two primary reasons:
1) If you are using RS232 then there can only be two devices on the cable segment.
2) If you are using RS485 then the 2nd client will not know to process the poll from the 1st client. It will cause errors.

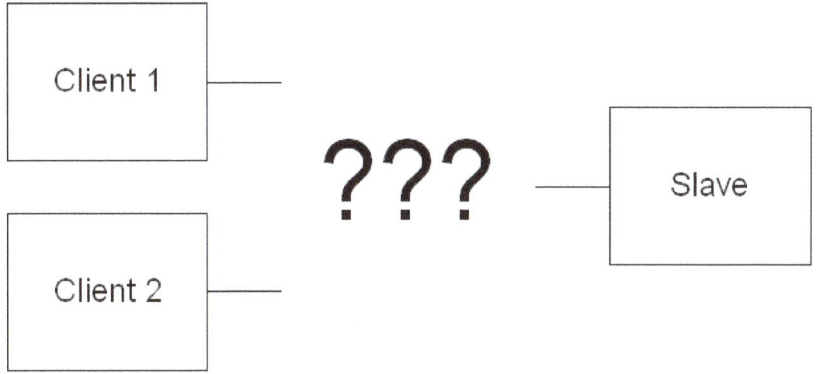

Figure 2: Using a FieldServer with an appropriate configuration solves this problem whether you are using RS232 or RS485.

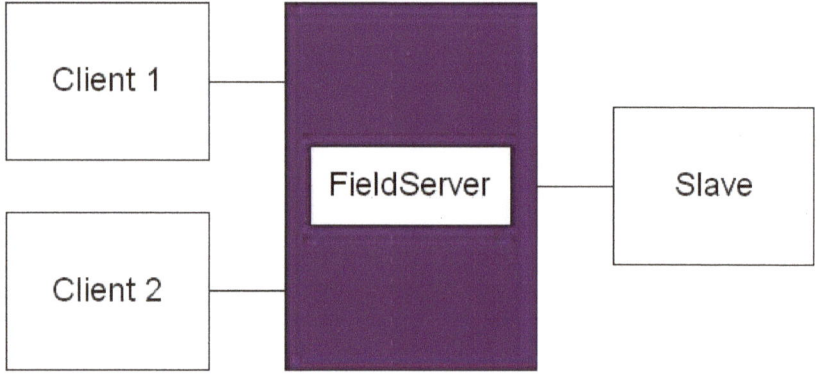

Figure 3: Each client is on its own port. Thus each client does not see poll messages from the other client. In this example client#1 sends a poll to the FieldServer. Then it is directed to a specific slave address. When the poll arrives at the FieldServer, the FieldServer checks the address against its configuration. If there is no match then an exception response is sent. If there is a match the FieldServer determines the port that the matching slave is configured on. The poll message is then relayed to the slave port.

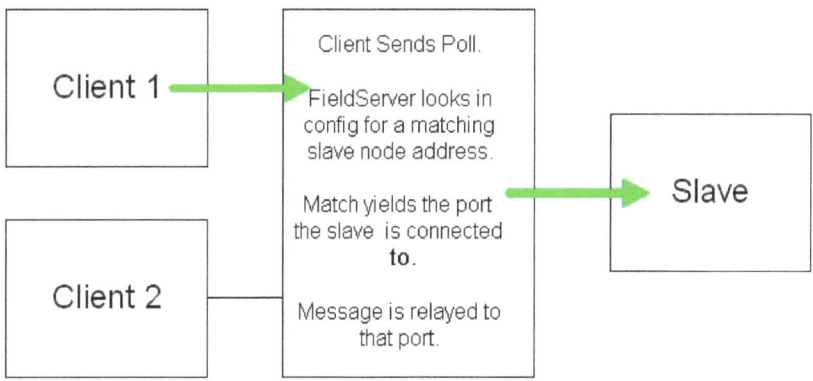

Figure 4: The slave responds. The FieldServer relays the response to client#1. The FieldServer also extracts the data from the response and stores in a temporary location (FieldServer calls that a cache block). The duration/expiry of the storage is configurable.

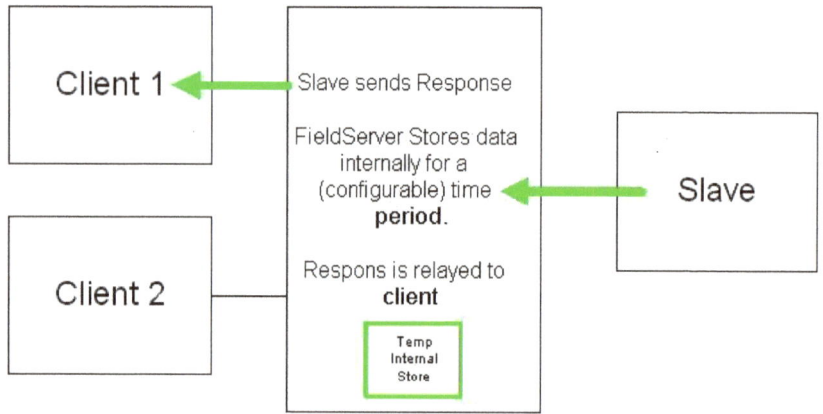

Figure 5: If any client requests the same data (client#1 or #2) and the data has not expired then the FieldServer responds with data from the temporary storage.

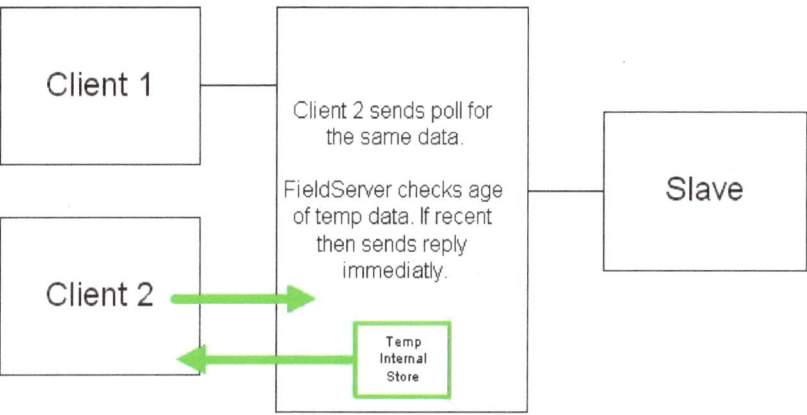

Figure 6: If any client requests different data or if the temporary data has expired then the match and relay process is repeated requesting the new data.

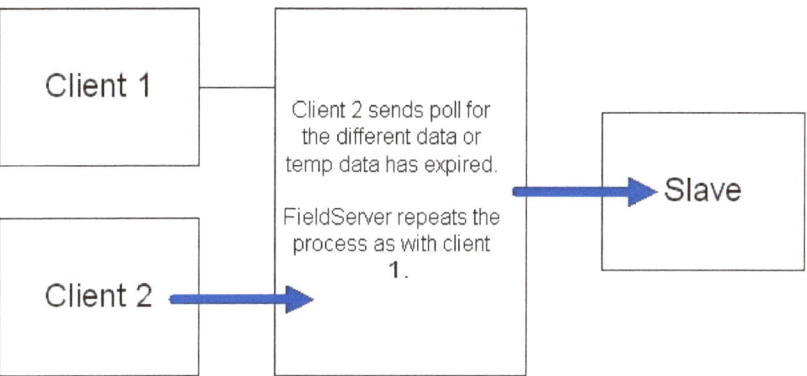

Figure 7: The slave responds, the response is relayed to the client doing the polling (Client#2 in this case) and the data is stored temporarily so that it is available to the other client.

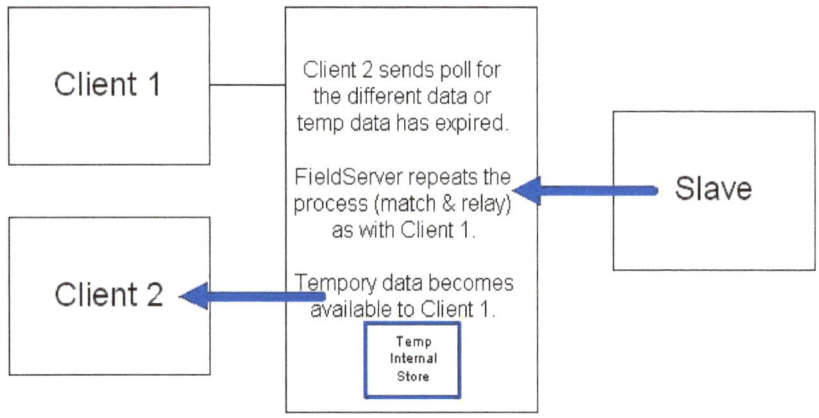

12. OLD DEVICE – SLOW PROCESSORS – LIMITED CAPABILITY

Many older devices have old microprocessors that can't do too much work at once. Often this microprocessor is used to run the device and handle the Modbus communication.

It is not uncommon to see device with the following limitations.

* You can only read one data point per message. I.e. length must be 1.
* You must have a delay between sending messages.

13. MODBUS ASCII, JBUS, ENRON AND OTHER VARIANTS

There are several variants of Modbus. They are not interoperable. I.e. A Modbus RTU master cannot read a Modbus ASCII field device.

ASCII – an attempt to make the Modbus message human readable but encoding the hex value of each byte in ascii. Stupid. Doubles the message length.

Jbus – Highway robbery. A Modbus RTU variation that allows more than 9999 of each data type to be read. These days most vendors include this in their RTU drivers so you don't have to pay extra.

Enron – Came up with a way of carrying other data in the Modbus messages. They used multiple words to form data objects. Essentially a set of conventions. Both the client and server must support them.

MODBUS RS232, RS485 AND TCP/IP

14. HOW MODBUS IS TRANSPORTED

There are 3 main physical layers for Modbus.

RS232 : One master and one slave. Typically a cable with 3 conductors with max length of approx a couple of hundred feet. Usually easy. Sometimes some jumpers are required at one end to defeat handshaking.

RS485: One master and up to 128 slaves but take care to read more if you plan on more than 32. There are two wiring systems – so called 2-wire and so called 4-wire. They can be incompatible but usually 4-wire devices can be made to work on 2-wire systems. Each device must have a unique address and all devices must be set to the same baud rate, data bits, stop bits and parity. Usually easy to implement. The RS485 physical layer allows up to 128 devices to be installed on a single network with a max physical length of 4000ft and speeds up to 115k baud. Using repeaters allows the length to be increased. Compare to Ethernet where the spec allows a max of 100 meters (330ft) on a single unrepeated segment.

TCP/IP: All devices are essentially peers. A single device can be a master and a server. Routers can be used to connect sub-nets together. Broadcasts are almost ever used so are not an issue.

15. MODBUS ON RS232

RS232 requires a minimum of 3 conductors to connect the two devices. Rx, Tx and Ground.

Some devices implement hardware handshaking. This means that before they send a message some voltage must be applied to one of the other pins on the port. If hardware handshaking is active on the device, then you will never get a response until you bypass it or implement it. We recommend bypassing it because there are often differences in the ways that vendors implemented it.

Here are typical jumper schemes that can be applied to defeat handshaking.

Connect these pins together on the 9 Pin D-Type connector connected to the server device.

Pin	Function
'1	DCD
4	DTR
6	DSR

And

Pin	Function
'7	RTS
'8	CTS
'9	RI (often omitted)

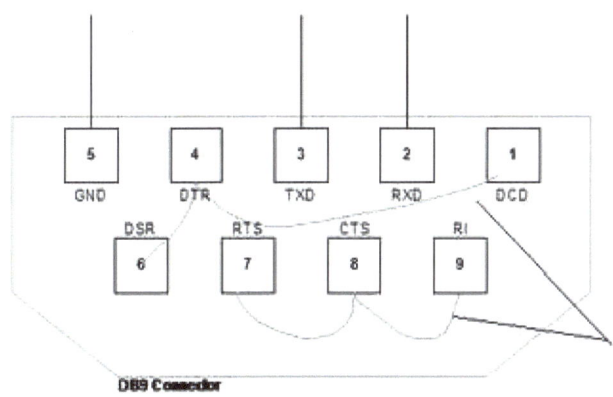

DB9 Connector

Chipkin
Automation Systems

16. MODBUS ON RS485

Search the Internet on RS485 you will find Bob Perrins's article called the "THE **ART** AND SCIENCE OF **RS-485**". It is his reference to Art that makes RS485 bad. What he means is that RS485 is often non-trivial and getting a network working can rely more on experience and experimentation.

Here is our simplified advice :

 Tip #1 – 3 Wires not 2

RS485 is a 3 conductor network. You take a huge risk by not installing the 3rd conductor. You risk blowing 485 ports, you risk unstable operation (works sometimes and doesn't work other times) and finally you risk re-installation. For a more detailed discussion read this article http://www.chipkin.com/articles/rs485-cables-why-you-need-3-wires-for-2-two-wire-rs485. The more power sources used to power devices, the greater the physical separation of devices, the less well-grounded devices and power sources are the greater the risk. Remember this statement: The so called Ground Terminal on a RS485 interface is not a connection to ground. It is a common reference signal. The voltage level on the Tx/Rx conductors are measured **relative** to this voltage level.

You can (if you must) use a shield drain wire as the 3rd conductor (ground reference conductor).

 Tip #2 – Connection Order

Always connect the ground reference conductor first if you are connecting a device that is powered up or you are connecting your laptop an operating network.

OR

Always choose devices that have optical isolation - this almost always will protect the RS485 transmitter / receivers.

 Tip #3 – Shield

You can get away without the shield. The twisted pair used for Tx and Rx is more effective at noise cancellation than the shield.

 Tip #4 – Cable Location

Take care where you run your cables. It seems obvious not to wind your cable around other cables or sources of electricity / magnetism. People are often surprised to find that the worst source of induced noise are switching DC loads. Another big culprit are Variable Frequency drives.

Tip Advice #5 – Cable Type

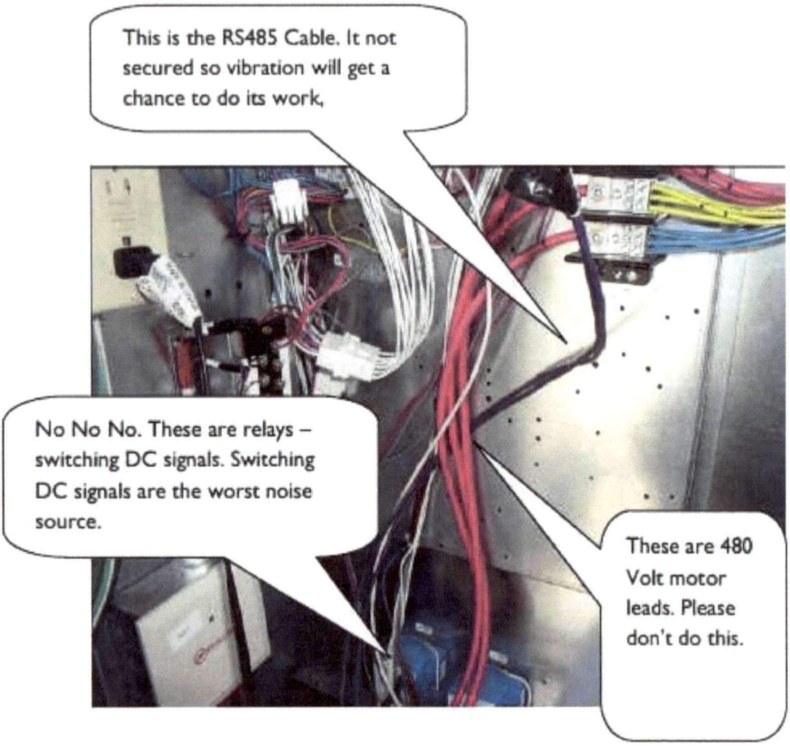

Cable selection does make a difference.

All cables offer impedance (resistance). Some cables are designed so that the impedance is relatively independent of distance. You want one of these cables. A clue to knowing if you selected one is to look at the cable's Nominal Impedance. If they quote a number such a 100Ohms you have a good cable. If they quote an impedance per meter/foot you have chosen the wrong kind.

Wrong in the sense – to determine the value of terminating resistors now requires measurements and calculations. Choose low capacitance cables.

Can you use Cat5 cable? Yes. Use one pair for Tx,Rx and a conductor from another pair for the ground reference signal.

We recommend these two cables.

Belden 3106A

Multi-Conductor – EIA Industrial RS-485 PLTC/CM 22 AWG stranded (7x30) tinned copper conductors, Datalene® insulation, twisted pairs, overall Beldfoil® shield (100% coverage) plus a tinned copper braid (90% coverage), drain wire, UV resistant PVC jacket.

Belden 3107A

Multi-Conductor - EIA Industrial RS-485 PLTC/CM 22 AWG stranded (7x30) tinned copper conductors, Datalene® insulation, twisted pairs, overall Beldfoil® shield (100% coverage) plus a tinned copper braid (90% coverage), drain wire, UV resistant PVC jacket.

 Tip #6 – Number of Devices per Trunk

How do you put more than 32 devices on a single RS485 trunk?

The simple answer is use a repeater but in practice one isn't always necessary.

The RS485 standard is based on 32 devices. Since the standard was developed most RS485 chips present less than the full unit load originally specified. Today you get half and quarter load devices. Thus to see how many devices you can install you simply get the data sheets and add the loads. Look for "UL" on the data sheet. It stands for Unit Load.

 Tip #7 – Cable Length

Cable Lengths and Baud Rates

Practically speaking you can go up to 4000 feet at baud rates up to 76800 baud. Above that you need to do a little math and reduce the length. For example, at 115k baud your cable should not be much longer than 2500 feet.

However, the higher the baud rate the more sensitive the cable is to the quality of installation – issues like how much twisted pair is unwound at each termination start to become very very important.

Our advice: For longer networks with lots of devices, choose 38k400 baud over 76k800 baud and optimize using COV, separate networks and by setting the Max Master to a lower number.

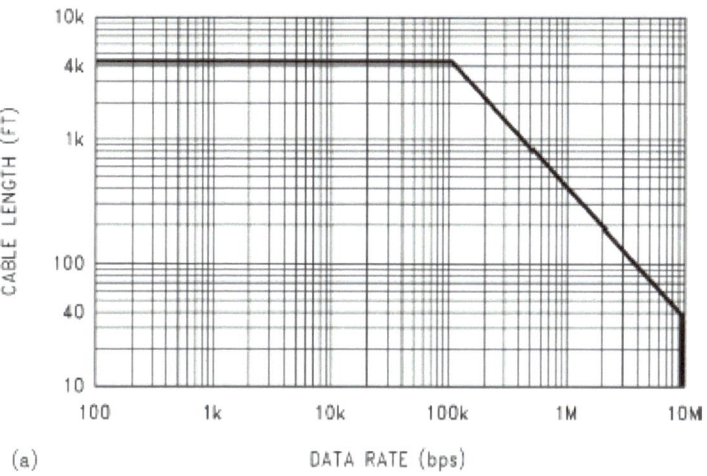

(a)   DATA RATE (bps)

*Source: **Ten Ways to Bulletproof RS-485 Interfaces** National Semiconductor*

Application Note 1057 John Goldie October 1996

 Tip #8 – Bandwidth Issues

How many devices to install on a single RS485 Trunk (Bandwidth Issues).

There are non-electrical considerations to determine how many devices you put on a Modbus trunk network. It's not possible to provide a calculator to work out how many devices to install on a single network but the following list provides some help in assessing bandwidth considerations.

Consider the following factors.

- A single Modbus message can only read consecutive data points.

  If you need to read 40001 and 40003 you must either:

  read 40001 length=3

  read 40001 length=1 and read 40003 length =3 (2 messages and 2 responses)

- A single Modbus message cannot read more than 125 16 bit words.

  The more dispersed the Modbus points you are reading the more messages and responses you will need.

  For example. If you need to read 40001 and 40128 then you will need at least two messages because all the data cannot be read in one message.

  Some devices have more severe limits. For example Crestron can only read 8 registers at a time.

- A single Modbus message can only read data of one type.

  If you need to read a coil and a holding register you will need at last one message for each.

- There may be some latency in the server devices – a time it takes to respond to messages.

  Some devices take up to 1 second between receiving a message and responding.

  Some devices can only be polled once per x seconds.

- What is the baud rate?

  Divide the baud rate by 10 to get approx characters per second.

  Divide the result by 2 to get approx number of words per second.

Thus at 19200 baud it takes approx to read 125 registers.

Poll = 10 bytes at 1920 per sec

Server latency

Response = 125 words at 960 per sec.

Client Latency (delay in storing response and sending next)

Approx 0.15 secs to .35 secs with typical latencies.

 Tip #9 – What Can Go Wrong

What can go wrong with 485?

Let's say you adopted all the best practices for installation of the network but you get intermittent or unacceptable performance because of packet loss, noise, collisions … Then you should consider hiring an expert to resolve your problems because now you are in the 'Art' part of RS485. These are some of the things they will look at.

● Reflections.

Without a scope and expertise you won't know this is a factor. It is easy and cheap to eliminate. Look at the cable spec. Find the nominal impedance. Buy two resistors of the same value. At each end of the trunk install the resistors between the Tx and Rx terminals. If you don't have obvious ends of the trunk (because you created a star) then we recommend re-cabling to form a linear trunk or we wish you luck.

Some devices have terminating resistors built into them. If the vendor did a poor job, the default is to have the resistor active and they must be disabled unless they are the terminating devices on the network. Read vendor doc.

● Biasing, Idle State Biasing, Fail Safe Biasing, Anti Aliasing

There are a whole string of terms uses as synonyms to describe this phenomenon.

To use two wires (as opposed to full duplex 4 wire) for RS485 each device's transmitter and receiver must be set to an idle state to release the line for others use. Releasing the line means allowing it to 'float'. It must not be allowed to float at any voltage level so devices have pull up/down resistors to pull the line to an allowable 'floating' voltage (the floating state is also known as the tri-state.) The load presented by other devices on the network affects this floating so the resistor values may need to be changed depending on the number of devices installed and the values of the pull up/down resistors they are using. (You can imagine how tricky it is going to be to resolve this). If a device floats out of the specified range then to other devices it will look like the floating device isn't floating at all. The other devices will think that it is transmitting or receiving and thus blocking the line.

The simplest way of knowing if this is a factor – Does the device work properly when it is the only device on the network? When you install it in the full network other devices or this device stops working properly. This device and/or the pull up/down resistors of other devices are candidates for investigation.

A number of vendors have a range of pull up/down resistors installed and allow you to change the selection using software or jumpers.

● Line Drive On / Off

To use two wires for RS485 each device's transmitter and receiver must be set to an idle state to release the line for others use. When a device wants to send, it must grab the line. When it has finished sending, it must release the line. You can see there are potential problems here. What happens if one device waits too long after sending its last bit before releasing the line – it's possible that the other devices will miss some bits of data.

 Tip #10 – Topology

Take care with the topology. The best topology is a single trunk that in-outs on the terminal blocks of each device it connects. What do we mean by best? We mean the choice which is least likely to cause problems.

Best arrangement. (Showing TX conductor for reference only)

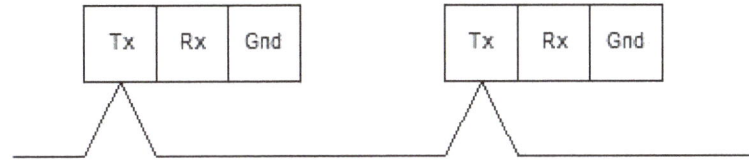

Getting worse. Making the connections to the RS485 terminals, drops instead of connections starts to give the electrical signals all kinds of complicated paths for reflections and harmonics. It is obvious that if the drops are long and are not twisted then you also have more chance to induce noise. (Showing TX conductor for reference only)

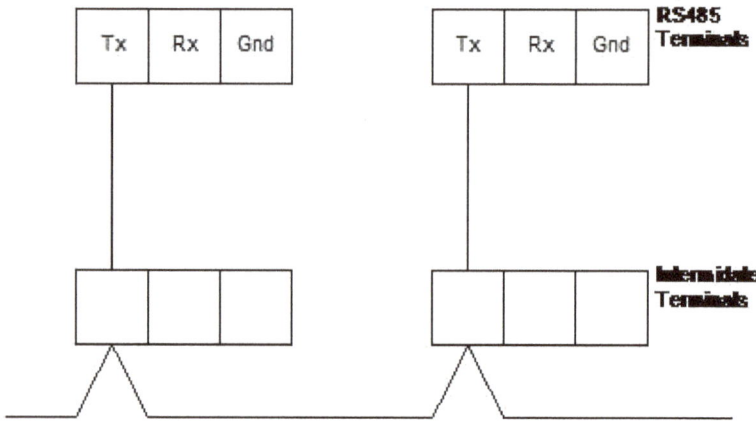

Worst. Avoid Star configurations. They are so much harder to debug when it gets tricky. (Showing TX conductor for reference only)

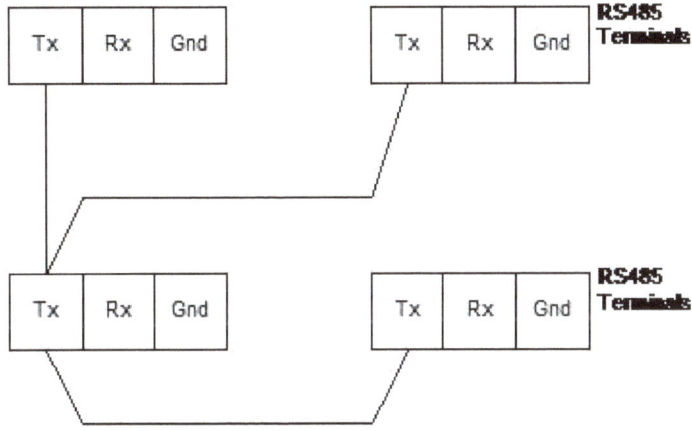

MODBUS RESOURCES, TESTING AND TROUBLE SHOOTING

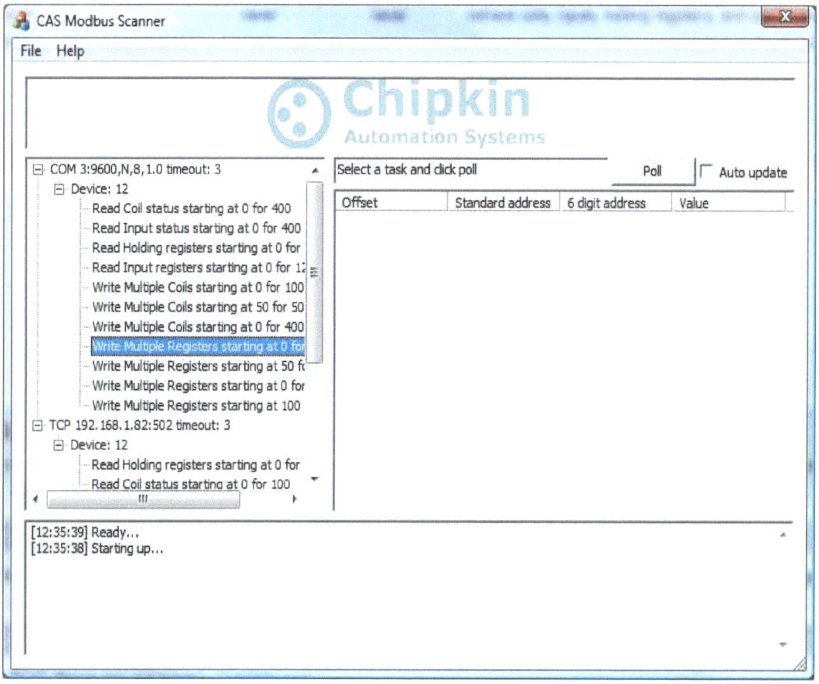

CAS Modbus Scanner

File   Help

Chipkin
Automation Systems

Select a task and click poll   Poll   ☐ Auto update

⊟ COM 3:9600,N,8,1.0 timeout: 3
  ⊟ Device: 12
    ─ Read Coil status starting at 0 for 400
    ─ Read Input status starting at 0 for 400
    ─ Read Holding registers starting at 0 for
    ─ Read Input registers starting at 0 for 12
    ─ Write Multiple Coils starting at 0 for 100
    ─ Write Multiple Coils starting at 50 for 50
    ─ Write Multiple Coils starting at 0 for 400
    ─ Write Multiple Registers starting at 0 for
    ─ Write Multiple Registers starting at 50 fo
    ─ Write Multiple Registers starting at 0 for
    ─ Write Multiple Registers starting at 100
⊟ TCP 192.168.1.82:502 timeout: 3
  ⊟ Device: 12
    ─ Read Holding registers starting at 0 for
    ─ Read Coil status starting at 0 for 100

Offset	Standard address	6 digit address	Value

[12:35:39] Ready...
[12:35:38] Starting up...

17. WHAT TO TAKE TO SITE WITH YOU

Here is a list of tools and resources you should carry with you to site for Modbus commissioning.

1. USB->485 converter

CAS uses: USB-COMI-SI-M from

http://www.usbgear.com/computer_cable_details.cfm?sku=USB-COMi-SI-M&cats=476&catid=494,476,199,461,106,1009,601

2. USB->232 Converter

Any will do

3. Laptop

4. Wireshark packet sniffer software – free download

http://www.wireshark.org/download.html

5. CAS Modbus Scanner – free download

CAS Modbus Scanner is a utility to retrieve coils, inputs, holding registers, and input registers from a Modbus enabled device. Values retrieved from the device can be viewed in many different formats including Binary, HEX, Uint16, Int16, Uint32, Int32, and Float32.
http://www.chipkin.com/cas-modbus-scanner

6. CAS Modbus Parser – free download

Have you ever needed to analyze a Modbus RTU message? The CAS Modbus RTU parser can analyze a Modbus RTU message and tell you if there are any errors in the message, what type of messages it is, what data is being written or read from your device, what device the message came from, and more...

http://www.chipkin.com/cas-modbus-rtu-parser/

7. Serial Break out box

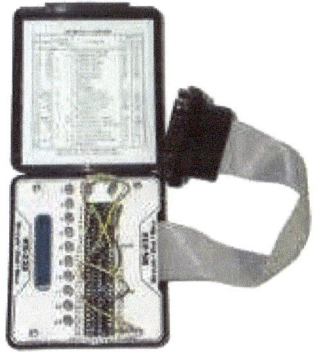

8. Lcom DB9-Terminal blocks (or similar)

http://www.l-com.com/item.aspx?id=8141 Male and Female

9. DB9 and DB25 male and female connector make-up kits (Solder free)

10. Rx / TX cross over.

It is useful to be able to swap the conductors connected to pins 2 and 3. Take a module with you. It is easier than changing the wires.

EG. Ziotek Null Modem Adapter DB25
http://www.cyberguys.com/product-details/?productid=751&rtn=750&core_cross=SEARCH_DETAIL_SIMILAR#page=page-1

female shown

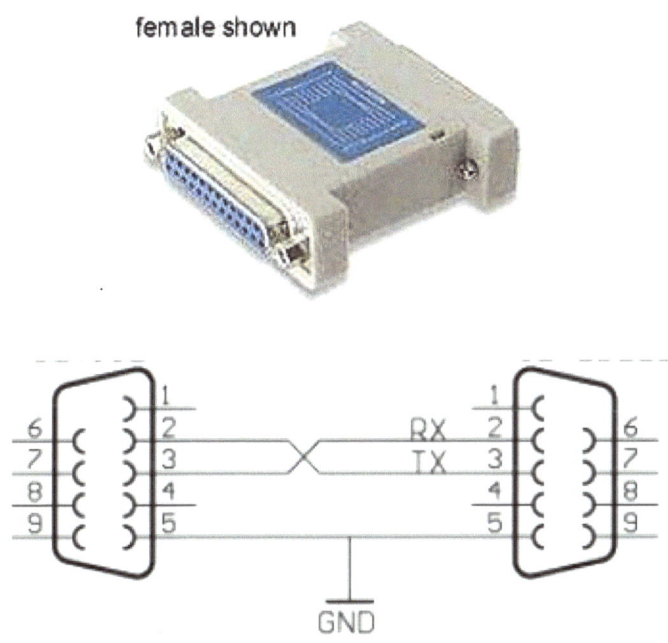

11. Terminating Resistors

Take 52.3, 75, 100, 120 and 150 Ohm resistors with you. 0.25 Watt is usually more than enough.

12. Gender Benders

13. Ethernet Patch cables

14. Hub

A hub is not a switch. A hub can be used for trouble shooting whereas only as 'supervised' switch can. Most switches are not supervised.

For more information read Appendix 5.

Appendix

18. TROUBLE SHOOTING MODBUS TCP/IP

REQUIRED TOOLS

Hub or Supervised Switch

Wireshark – Free Download

      http://www.wireshark.org/download.html

 Tip #1: You might not capture the traffic if you don't use a hub. Read the article on hub and switches to understand why.

 Tip #2: You can select the packets you capture to reduce log file size by defining a capture filter before you start the capture.

 We suggest you to avoid this. If you are short of space you can select which packets you save.

 Tip #3: You can select which packets you view from the total log by defining a display filter.

 Tip #4: You can select which packets to save in the log files.

 Tip #5: You can search for particular packets.

HOW TO CAPTURE WITH WIRESHARK

1. Capture – Main Menu

2. Interfaces – On Capture Menu

a) You get a list of network adapters. Pick the one connected to the network of interest. It's probably not the wireless adapter. Most often it's the adapter with the packet count increasing.

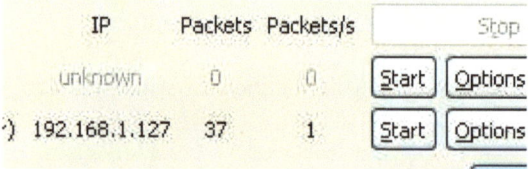

b) Select the Start button or

c) Select the options button to define a capture filter. Define the filter and click start.

3. A list of packets accumulates on the screen.

	Time	Source	Destination	Protocol	Info
213	302.317110	192.168.1.118	192.168.1.81	TCP	afs3-b
214	302.319551	192.168.1.81	192.168.1.118	TCP	asa-ap
215	302.319561	192.168.1.118	192.168.1.81	TCP	afs3-b
216	302.321023	192.168.1.118	192.168.1.81	Modbus/T	que
217	302.325145	192.168.1.81	192.168.1.118	TCP	asa-ap
218	302.328580	192.168.1.81	192.168.1.118	Modbus/T	respor
219	302.410712	fe80::d57f:df48:4ae3:	ff02::c	SSDP	M-SEAR
220	302.456205	192.168.1.118	192.168.1.81	TCP	afs3-b
221	302.459244	192.168.1.81	192.168.1.118	TCP	[TCP D
222	303.828843	Cisco-Li_3a:27:b8	AsustekC_b7:84:06	ARP	who ha
223	303.828852	AsustekC_b7:84:06	Cisco-Li_3a:27:b8	ARP	192.16
224	305.138317	192.168.1.118	192.168.1.81	TCP	afs3-b
225	305.139814	192.168.1.118	192.168.1.81	TCP	afs3-u
226	305.141845	192.168.1.81	192.168.1.118	TCP	asa-ap
227	305.141855	192.168.1.118	192.168.1.81	TCP	afs3-u
228	305.143343	192.168.1.118	192.168.1.81	Modbus/T	que

4. Apply a Display Filter. More on display filters later. For now simply type **mbtcp** into the filter field and click apply.

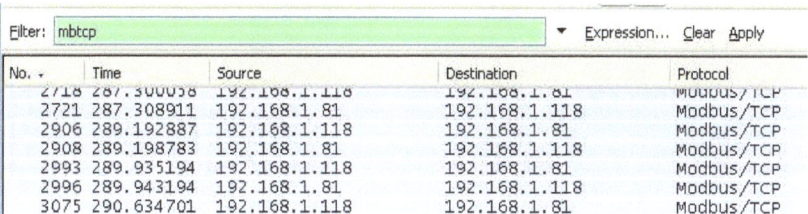

No. ▾	Time	Source	Destination	Protocol
2718	287.300058	192.168.1.118	192.168.1.81	Modbus/TCP
2721	287.308911	192.168.1.81	192.168.1.118	Modbus/TCP
2906	289.192887	192.168.1.118	192.168.1.81	Modbus/TCP
2908	289.198783	192.168.1.81	192.168.1.118	Modbus/TCP
2993	289.935194	192.168.1.118	192.168.1.81	Modbus/TCP
2996	289.943194	192.168.1.81	192.168.1.118	Modbus/TCP
3075	290.634701	192.168.1.118	192.168.1.81	Modbus/TCP

5. Find the packet you are interested in. Click on it to select it. A breakout of the selected packet's data is shown below the packet list.

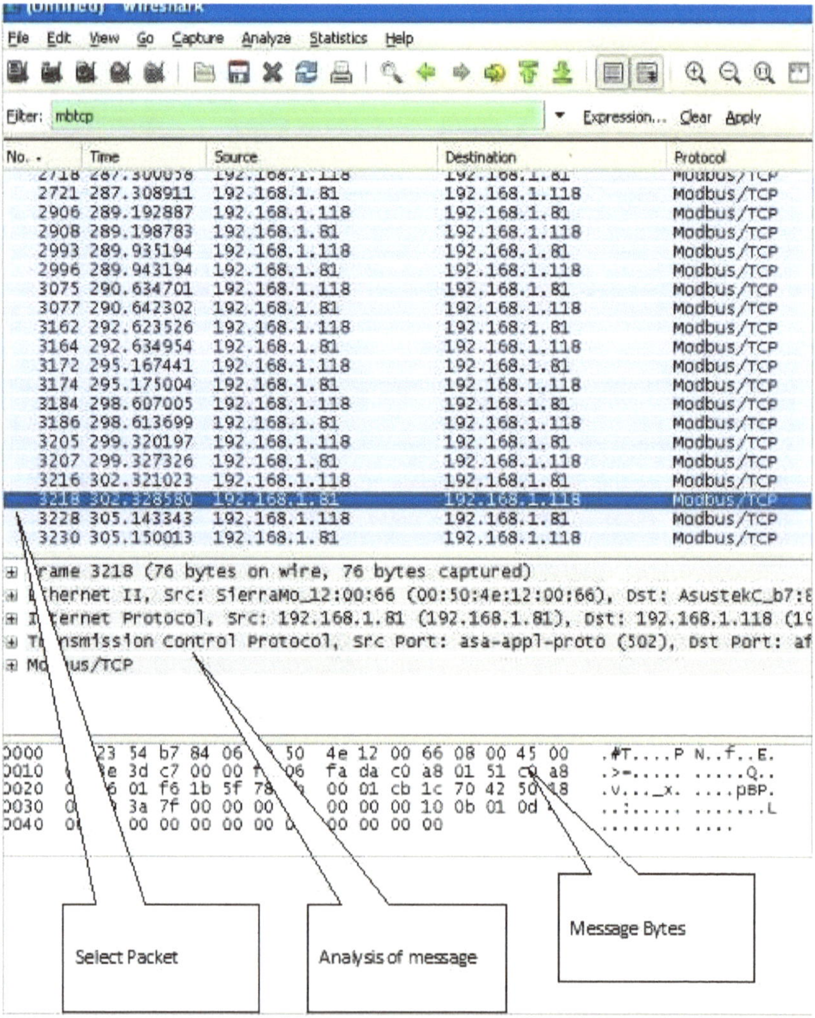

```
(Untitled) - Wireshark

File  Edit  View  Go  Capture  Analyze  Statistics  Help

Filter: mbtcp                                          ▼  Expression...  Clear  Apply

No. -   Time          Source            Destination        Protocol
2718  287.300058  192.168.1.118     192.168.1.81       Modbus/TCP
2721  287.308911  192.168.1.81      192.168.1.118      Modbus/TCP
2906  289.192887  192.168.1.118     192.168.1.81       Modbus/TCP
2908  289.198783  192.168.1.81      192.168.1.118      Modbus/TCP
2993  289.935194  192.168.1.118     192.168.1.81       Modbus/TCP
2996  289.943194  192.168.1.81      192.168.1.118      Modbus/TCP
3075  290.634701  192.168.1.118     192.168.1.81       Modbus/TCP
3077  290.642302  192.168.1.81      192.168.1.118      Modbus/TCP
3162  292.623526  192.168.1.118     192.168.1.81       Modbus/TCP
3164  292.634954  192.168.1.81      192.168.1.118      Modbus/TCP
3172  295.167441  192.168.1.118     192.168.1.81       Modbus/TCP
3174  295.175004  192.168.1.81      192.168.1.118      Modbus/TCP
3184  298.607005  192.168.1.118     192.168.1.81       Modbus/TCP
3186  298.613699  192.168.1.81      192.168.1.118      Modbus/TCP
3205  299.320197  192.168.1.118     192.168.1.81       Modbus/TCP
3207  299.327326  192.168.1.81      192.168.1.118      Modbus/TCP
3216  302.321023  192.168.1.118     192.168.1.81       Modbus/TCP
3218  302.328580  192.168.1.81      192.168.1.118      Modbus/TCP
3228  305.143343  192.168.1.118     192.168.1.81       Modbus/TCP
3230  305.150013  192.168.1.81      192.168.1.118      Modbus/TCP

⊞ Frame 3218 (76 bytes on wire, 76 bytes captured)
⊞ Ethernet II, Src: SierraMo_12:00:66 (00:50:4e:12:00:66), Dst: AsustekC_b7:8
⊞ Internet Protocol, Src: 192.168.1.81 (192.168.1.81), Dst: 192.168.1.118 (19
⊞ Transmission Control Protocol, Src Port: asa-appl-proto (502), Dst Port: af
⊞ Modbus/TCP

0000   23 54 b7 84 06   50 4e 12 00 66 08 00 45 00   .#T....P N..f..E.
0010   e 3d c7 00 00 f  06 fa da c0 a8 01 51 c0 a8   .>=..... .....Q..
0020   6 01 f6 1b 5f 78  00 01 cb 1c 70 42 50 8    .v..._X. ....pBP.
0030   3a 7f 00 00 00   00 00 00 10 0b 01 0d       ..:..... ........L
0040   00 00 00 00 00 0  00 00 00 00              ........ ....
```

Select Packet

Analysis of message

Message Bytes

6. You can break out the level of detail by expanding the sections of the packet.

Think of a Modbus packet as a letter you send to a Modbus device. When you take it to the Modbus post office, the clerk says he does not understand the address. He passes it to the TCP/IP clerk. The TCP/IP clerk takes your letter and puts it in a bigger envelope. He addresses the envelope with a TCP/IP address. He passes it to the Ethernet post office clerk. The Ethernet clerk takes your letter and puts it in a bigger envelope. He addresses the envelope with a hardware address and sends it to that computer. When it arrives the process is reversed until finally the contents are passed to the Modbus application.

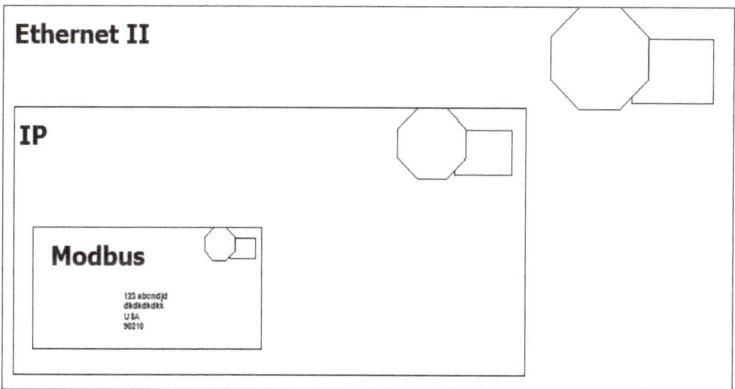

Ethernet packets contain packets from other higher level   protocols nested inside each other. You drill down to see the detail you want.

In the example below you can see the Modbus packet nested inside an IP protocol packet which is in turn nested inside an Ethernet packet.

7. Drill down to see the Modbus info

```
⊞ Frame 3218 (76 bytes on wire, 76 bytes captured)
⊞ Ethernet II, Src: SierraMo_12:00:66 (00:50:4e:12:00:66), Dst: AsustekC_b7:84:06 (00:23:54:b7:84:06)
⊞ Internet Protocol, Src: 192.168.1.81 (192.168.1.81), Dst: 192.168.1.118 (192.168.1.118)
⊞ Transmission Control Protocol, Src Port: asa-appl-proto (502), Dst Port: afs3-bos (7007), Seq: 1, Ack: 13, Len: 22
⊟ Modbus/TCP
     transaction identifier: 2
     protocol identifier: 0
     length: 16
     unit identifier: 11
  ⊟ Modbus
     function 1:  Read coils
     byte count: 13
     Data
```

Before you start a capture you can specify a capture filter. The effect of the filter is to prevent all packets being captured. Doing this can save space when you save the log and it might make it easier to find the packets you are interested in. However, there is some risk that you might filter out the packets of interest.

For example, a Modbus device might not operate correctly because it is being hammered with packets from another protocol being sent incorrectly to the Modbus device. Our advice is to capture as much as possible and then filter what is displayed.

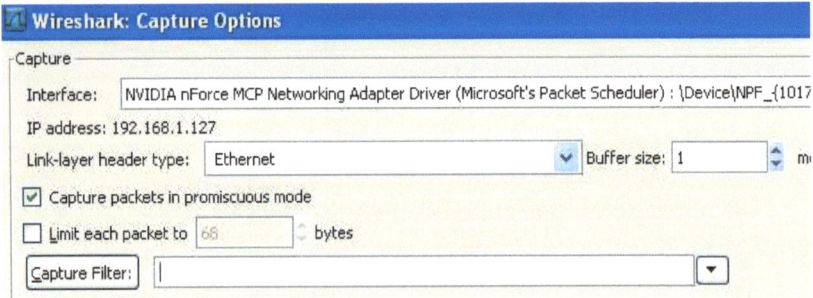

CAPTURE FILTERS

Here are some sample filters

Examples:

Capture only traffic to or from IP address 172.18.5.4:

> host 172.18.5.4

Capture only traffic to or from IP address 172.18.5.4 but exclude all FieldServer RUINET messages

> host 192.168.1.81 and port not 1024

Capture traffic to or from a range of IP addresses:

> net 192.168.0.0/24

or

> net 192.168.0.0 mask 255.255.255.0

Capture traffic from a range of IP addresses:

> src net 192.168.0.0/24

or

> src net 192.168.0.0 mask 255.255.255.0

Capture traffic to a range of IP addresses:

dst net 192.168.0.0/24

or

dst net 192.168.0.0 mask 255.255.255.0

Capture only Modbus traffic: Assumes every device is compliant and is using the standard port.

port 502

Useful Hint :

 It is easy to sort packets by source or destination IP, Click the column headings.

Useful Hint :

 You can mark, packets you find interesting. Then later you can save, display or print the marked packets.

DISPLAY FILTERING

 Useful Hint :

Any capture filter can be used as a display filter. You can use the expression builder to build selection criteria for filters.

SEARCHING

Looking for messages to/from particular devices

ip.addr == 192.168.1.90   Sent to/Sent From

ip.dst_host == "192.168.1.90"

ip.src_host == "192.168.1.90"

Looking for messages to/from particular Modbus devices

modbus_tcp.unit_id == 11 (look for all message which were sent to Modbus Device #11)

You can use the expression builder to build filter expressions

Filter: [   ]   Expression... Clear Apply

Expression Builder

From the drop down list of protocols there is one specifically related to Modbus. They are shown below

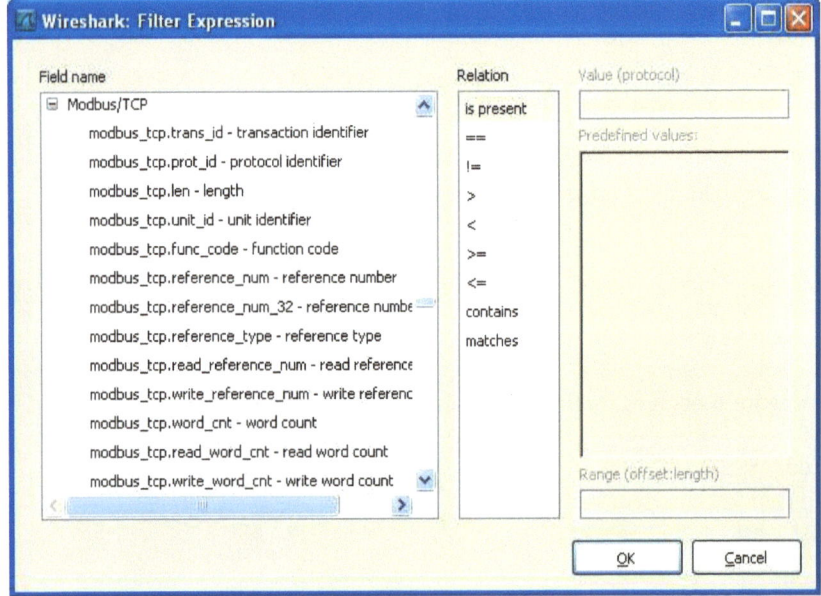

19. USING THE CAS MODBUS SCANNER

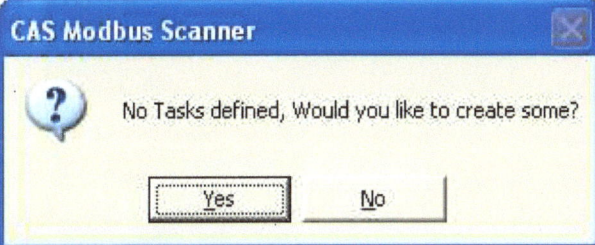

Add a connection

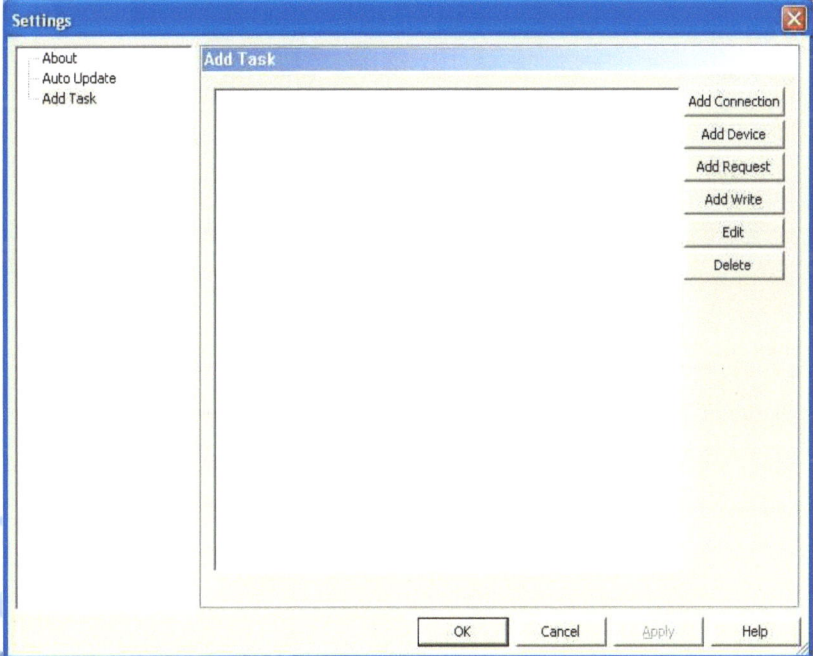

Choose from Serial or Ethernet

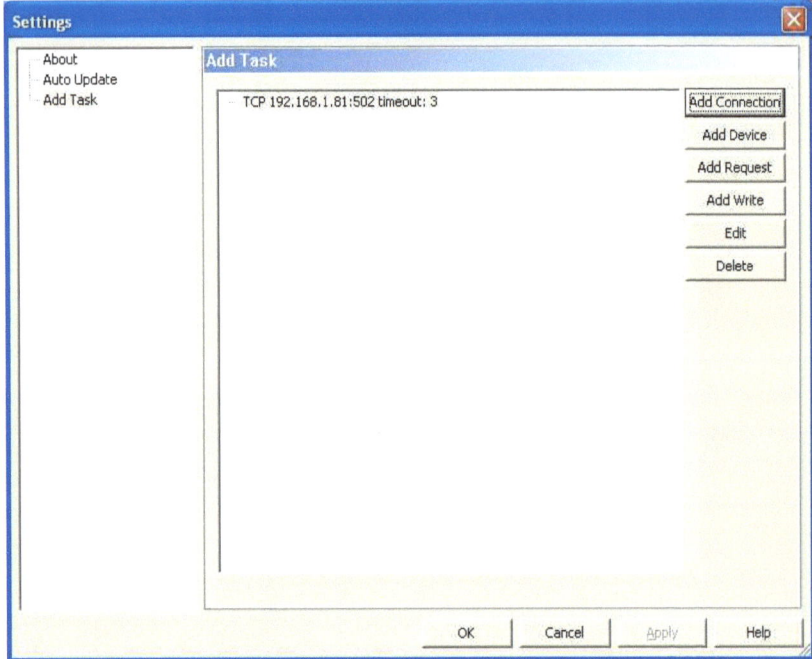

Add devices. The device number is the Modbus Device Number.

Add requests – polls for data. You can multiple requests for each device.

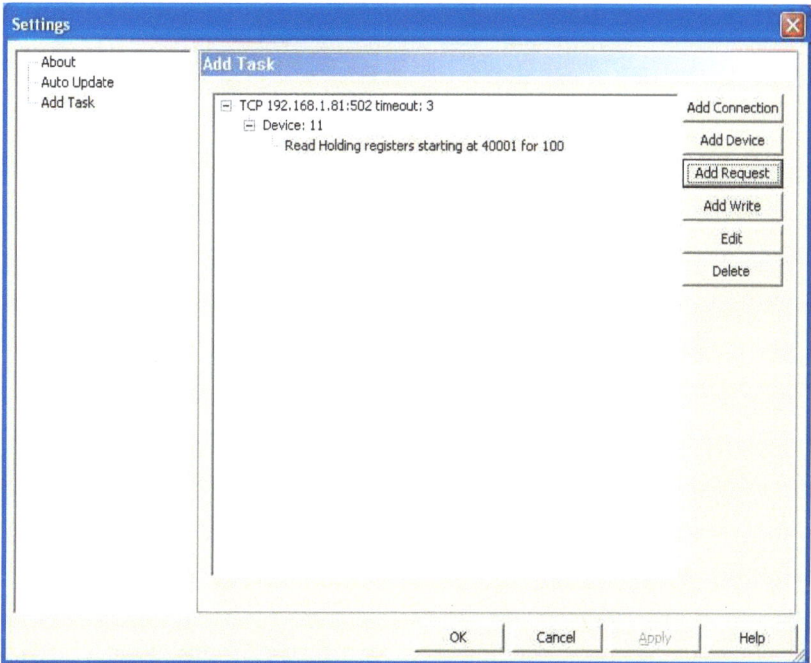

You can add multiple connections. More than one of each type. Each connection can have one or more device. Each device can have one or more requests.

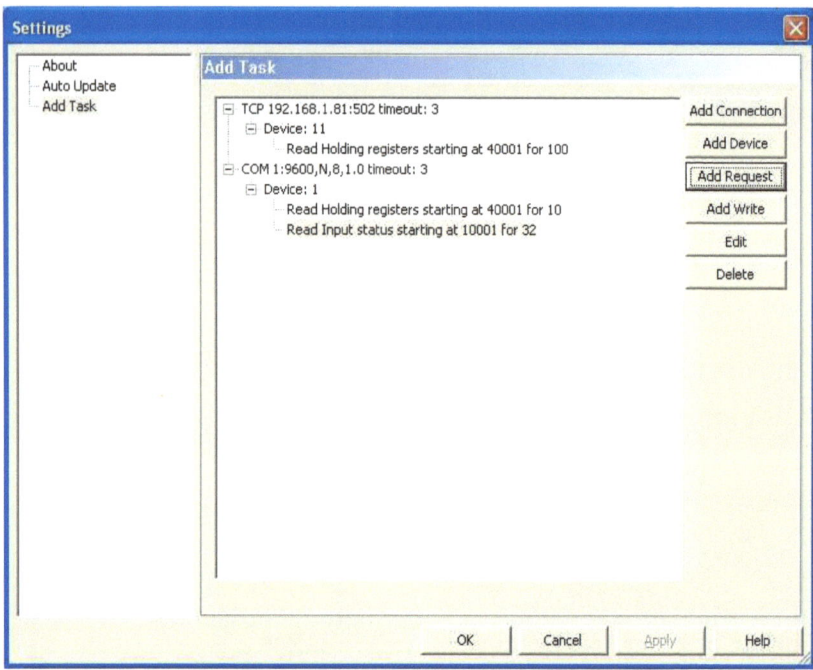

Once connections, devices and requests have been defined you can scan for data, exit or edit the settings.

To scan for data – Double click a request. It will be executed once. You can have the request auto repeat by checking the Auto Update box.

You will not get a response every time you poll. You may be polling the wrong device with the wrong IP address or wrong baud rate. You could be polling for points that don't exist ... there are many reasons. If you don't get a response this is called a timeout.

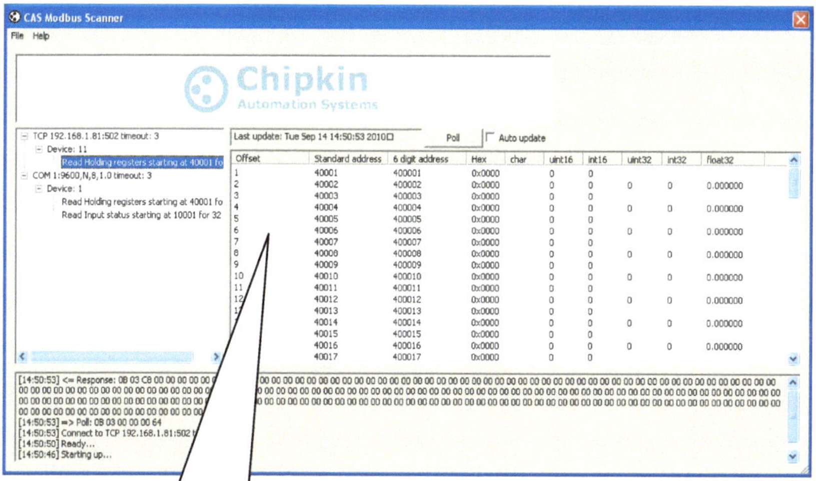

CAS Modbus Scanner

File   Help

Chipkin
Automation Systems

Offset	Standard address	6 digit address	Hex	char	uint16	int16	uint32	int32	float32
1	40001	400001	0x0000		0	0			
2	40002	400002	0x0000		0	0	0	0	0.000000
3	40003	400003	0x0000		0	0			
4	40004	400004	0x0000		0	0	0	0	0.000000
5	40005	400005	0x0000		0	0			
6	40006	400006	0x0000		0	0	0	0	0.000000
7	40007	400007	0x0000		0	0			
8	40008	400008	0x0000		0	0	0	0	0.000000
9	40009	400009	0x0000		0	0			
10	40010	400010	0x0000		0	0	0	0	0.000000
11	40011	400011	0x0000		0	0			
12	40012	400012	0x0000		0	0	0	0	0.000000
1	40013	400013	0x0000		0	0			
	40014	400014	0x0000		0	0	0	0	0.000000
	40015	400015	0x0000		0	0			
	40016	400016	0x0000		0	0	0	0	0.000000
	40017	400017	0x0000		0	0			

- TCP 192.168.1.81:502 timeout: 3
  - Device: 11
    - Read Holding registers starting at 40001 fo
- COM 1:9600,N,8,1.0 timeout: 3
  - Device: 1
    - Read Holding registers starting at 40001 fo
    - Read Input status starting at 10001 for 32

Last update: Tue Sep 14 14:50:53 2010☐   Poll   ☐ Auto update

[14:50:53] <= Response: 0B 03 C8 00 00 00 00 ...
[14:50:53] => Poll: 0B 03 00 00 00 64
[14:50:53] Connect to TCP 192.168.1.81:502 ...
[14:50:50] Ready...
[14:50:46] Starting up...

Data is displayed here in various formats.

When displayed as floats or 32 bit integers, the utility uses two consecutive registers to calculate the result. The value may not be what you expect because of byte/word order issues.

20. CONVERTING MODBUS 16 BIT NUMBERS TO 32 BIT NUMBERS

Often the Vendor documentation does not report the byte order in which registers are served or the order in which words must be combined to form 32 bit numbers. For this reason FieldServer provides 4 functions to convert Modbus 16 bit numbers to 32 bit numbers.

2.i16-1.i32

2.i16-1.i32-s

2.i16-1.i32-sw

2.i16-1.i32-sb

Each of these functions takes 2x 16 bits numbers to form a 32 bit number. Each processes the bytes in a different order.

 Practical Tip:

The easiest way to determine which function to use is to experiment. Look at the values in the FieldServer Data Arrays. If the values are obviously wrong try the other move functions. (Don't forget that some numbers may be scaled so the number you see in the Data Array may 10x or 100x too big / small).

Example:

In the move below if

DA_B01_REG[21] == 1 and

DA_B01_REG[22] == 2

Then

DA_B01_INT32[21] == 131073

Explanation

1 + 65536* 2 = 131073

Or In Hexadecimal

0×0001 + 0×0002 * 0×10000 = 0×20001

```
Moves
Source_Data_Array ,Source_Offset ,Target_Data_Array ,Target_Offset ,Length
,Function
DA_B01_REG ,21 ,DA_B01_INT32 ,21 ,2 ,2.i16-1.i32
```

By the way, the length is 2 because the move must process two values from the source.

Example:

DA_B01_REG[21] == 1 and

DA_B01_REG[22] == 2

Function In Use	Value found in DA_B01_INT32[21]
2.i16-1.i32	131073
2.i16-1.i32s	16777728
2.i16-1.i32-sw	65538
2.i16-1.i32-sb	33554688

21. HOW REAL (FLOATING POINT) AND 32-BIT DATA IS ENCODED IN MODBUS RTU MESSAGES

The article discusses some of the typical difficulties encountered when handling 32-bit data types via Modbus RTU and offers practical help for solving these problems.

The point-to-point Modbus protocol is a popular choice for RTU communications if for no other reason that it's basic convenience. The protocol itself controls the interactions of each device on a Modbus network, how device establishes a known address, how each device recognizes its messages and how basic information is extracted from the data. In essence, the protocol is the foundation of the entire Modbus network.

Such convenience does not come without some complications however, and Modbus RTU Message protocol is no exception. The protocol itself was designed based on devices with a 16-bit register length. Consequently, special considerations were required when implementing 32-bit data elements. This implementation settled on using two consecutive 16-bit registers to represent 32 bits of data or essentially 4 bytes of data. It is within these 4 bytes of data that single-precision floating point data can be encoded into a Modbus RTU message.

THE IMPORTANCE OF BYTE ORDER

Modbus itself does not define a floating point data type but it is widely accepted that it implements 32-bit floating point data using the IEEE-754 standard. However, the IEEE standard has no clear cut definition of byte order of the data payload. Therefore the most important consideration when dealing with 32-bit data is that data is addressed in the proper order.

For example, the number 123456.00 as defined in the IEEE 754 standard for single-precision 32-bit floating point numbers appears as follows:

Decimal	Hexidecimal Value				+/-	Exponent	Significand
	A	B	C	D			
123456.00	00	20	F1	47	0	10001111	1 .1110001001000000000000000

The affects of various byte orderings are significant. For example, ordering the 4 bytes of data that represent 123456.00 in a "B A D C" sequence in known as a "byte swap". When interpreted as an IEEE 744 floating point data type, the result is quite different:

Decimal	Hexidecimal Value				+/-	Exponent	Significand
	B	A	D	C			
-9.85402689122801930E+029	20	00	47	F1	1	11100010	1 .10001110000000000100000

Ordering the same bytes in a "C D A B" sequence is known as a "word swap". Again, the results differ drastically from the original value of 123456.00:

Decimal	Hexidecimal Value				+/-	Exponent	Significand
	C	D	A	B			
+1.08658251394509650E-019	F1	47	00	20	0	01000000	1 .00000000100011111110001

Furthermore, both a "byte swap" and a "word swap" would essentially reverse the sequence of the bytes altogether to produce yet another result:

Decimal	Hexidecimal Value				+/-	Exponent	Significand
	D	C	B	A			
+1.257363081x2E-127	47	F1	20	00	0	00000000	1 .0100000111110001010100110

Clearly, when using network protocols such as Modbus, strict attention must be paid to how bytes of memory are ordered when they are transmitted, also known as the 'byte order'.

DETERMINING BYTE ORDER

The Modbus protocol itself is declared as a 'big-Endian' protocol, as per the Modbus Application Protocol Specification, V1.1.b:

"Modbus uses a "big-Endian" representation for addresses and data items. This means that when a numerical quantity larger than a single byte is transmitted, the most significant byte is sent first."

Big-Endian is the most commonly used format for network protocols – so common, in fact, that it is also referred to as 'network order'.

Given that the Modbus RTU message protocol is big-Endian, in order to successfully exchange a 32-bit datatype via a Modbus RTU message, the endianness of both the master and the slave must be considered. Many RTU master and slave devices allow specific selection of byte order particularly in the case of software-simulated units. One must merely insure that both all units are set to the same byte order.

As a rule of thumb, the family of a device's microprocessor determines its endianness. Typically, the big-Endian style (the high-order byte is stored first, followed by the low-order byte) is generally found in CPUs designed with a Motorola processor. The little-Endian style (the low-order byte is stored first, followed by the high-order byte) is generally found in CPUs using the Intel architecture. It is a matter of personal perspective as to which style is considered 'backwards'.

If, however, byte order and endianness is not a configurable option, you will have to determine how to interpret the byte. This can be done requesting a known floating-point value from the slave. If an impossible value is returned, i.e. a number with a double-digit exponent or such, the byte ordering will most likely need modification.

PRACTICAL HELP

The FieldServer Modbus RTU drivers offer several function moves that handle 32-bit integers and 32-bit float values. More importantly, these function moves consider all different forms of byte sequencing. The following table shows the FieldServer function moves that copy two adjacent 16-bit registers to a 32-bit integer value.

Function Keyword	Swap Mode	Source Bytes	Target Bytes
2.i16-1.i32	N/A	[a b] [c d]	[a b c d]
2.i16-1.i32-s	byte and word swap	[a b] [c d]	[d c b a]
2.i16-1.i32-sb	byte swap	[a b] [c d]	[b a d c]
2.i16-1.i32-sw	word swap	[a b] [c d]	[c d a b]

The following table shows the FieldServer function moves that copy two adjacent 16-bit registers to a 32-bit floating point value:

Function Keyword	Swap Mode	Source Bytes	Target Bytes
2.i16-1.ifloat	N/A	[a b] [c d]	[a b c d]
2.i16-1.ifloat-s	byte and word swap	[a b] [c d]	[d c b a]
2.i16-1.ifloat-sb	byte swap	[a b] [c d]	[b a d c]
2.i16-1.ifloat-sw	word swap	[a b] [c d]	[c d a b]

The following table shows the FieldServer function moves that copy a single 32-bit floating point value to two adjacent 16-bit registers:

Function Keyword	Swap Mode	Source Bytes	Target Bytes
1.float-2.i16	N/A	[a b c d]	[a b][c d]
1.float-2.i16-s	byte and word swap	[a b c d]	[d c][b a]
1.float-2.i16-sb	byte swap	[a b c d]	[b a][d c]
1.float-2.i16-sw	word swap	[a b c d]	[c d][a b]

Given the vairous FieldServer function moves, the correct handling of 32-bit data is dependent on choosing the proper one. Observe the following behavior of these FieldServer function moves on the known single-precision decimal float value of 123456.00:

16-bit Values	Function Move	Result	Function Move	Result
0×2000 0x47F1	2.i16-1.float	123456.00	1.float-2.i16	0×2000 0x47F1
0xF147 0×0020	2.i16-1.float-s	123456.00	1.float-2.i16-s	0xF147 0X0020
0×0020 0xF147	2.i16-1.float-sb	123456.00	1.float-2.i16-sb	0×0020 0xF147
0x47F1 0×2000	2.i16-1.float-sw	123456.00	1.float-2.i16-sw	0x47F1 0×2000

Modbus for Field Technicians

Notice that different byte and word orderings require the use of the appropriate FieldServer function move. Once the proper function move is selected, the data can be converted in both directions.

Of the many hex-to-floating point converters and calculators that are available in the Internet, very few actually allow manipulation of the byte and word orders. One such utility is located at www.61131.com/download.htm where both Linux and Windows versions of the utilities can be downloaded.

Once installed, the utility is run as an executable with a single dialog interface. The utility presents the decimal float value of 123456.00 as follows:

One can then swap bytes and/or words to analyze what potential endianness issues may exist between Modbus RTU master and slave devices.

22. HUBS VS SWITCHES – USING WIRESHARK TO SNIFF NETWORK PACKETS

Gotcha #1 : Use a hub not a switch

Why: Switches don't copy all messages to all ports. They try and optimize traffic so when they learn which port a device is connected to they send all messages intended for that device to that port and stop copying to all ports. (The jargon they use for this function is 'learning mode')

How do you know it's a hub: Just because it calls itself a hub doesn't mean it is one.

- If it says full-duplex in the product description it's probably not a hub.

- A switch that allows you to turn off the learning mode is effectively a hub.

- A switch with a monitored port copies all messages to the monitored port and thus you can use that port as if it were a hub.

- If it says 'switch' and you can't turn off learning mode and it doesn't have a monitor port then it is not a hub.

- A router is never a hub.

Gotcha #2 : Mixing 10 and 100 mbits/sec can cause problems.

Not all hubs copy 10mbit messages to 100mbit ports and vice versa. Use a 10mbit/sec hub if you are on a mixed network – almost all other faster devices are speed sensing and will downgrade themselves to 10mbits/sec and thus you will see all the packets. This is not true of some building automation engines where the speed of the port is configured.

You can work around this problem by connecting higher speed devices to a self sending switch/hub and then connect that switch/hub to the 10mbit hub.

Recommended Hubs

- •10Mbit/sec Networks – DX-EHB4 – 4 Port 10 Mbps HUB
- •Netgear – DS104 Dual Speed HUB
- •10Mbit/sec Networks – D-LINK DE-805TP

www.ingramcontent.com/pod-product-compliance
Lightning Source LLC
Chambersburg PA
CBHW040832180526
45159CB00001B/163